ACCLAIM FOR THE ARCTIC CYCLE

Chantal Bilodeau is giving an artistic voice to the challenge of a changing Arctic environment.

—*Canadian Wildlife*

Bilodeau's Arctic Cycle gives us an ambitious and timely resource to explore the interconnected nature of people and the planet ... She has crafted tapestries that explore climate change in ways that humanize the science and recognize the complexities of competing interests. The Arctic Cycle is about more than climate change; it explores the commonality and diversity of humanity.

—Michael Mehler
ASSOCIATE PROFESSOR AND CHAIR OF COMMUNICATION, FILM, AND THEATRE, ALLEGHENY COLLEGE

Bilodeau's eight-part Arctic Cycle harnesses an artistic hallmark of serial storytelling – its ability to create an epic through self-sufficient yet interrelated components – to overcome a central challenge of climate change drama. It promises to bring a phenomenon of overwhelming proportions and import into meaningful cognitive and emotive focus ... The intricate aesthetic microcosms of the first two dramas – *Sila* and *Forward* – illustrate how acceptance of the Anthropocene requires a willingness to break with established Western notions of linearity and progress.

—Nassim W. Balestrini
PROFESSOR OF AMERICAN STUDIES AND INTERMEDIALITY, UNIVERSITY OF GRAZ, AUSTRIA

ACCLAIM FOR SILA,
THE FIRST PLAY IN THE ARCTIC CYCLE

Simply sublime ... *Sila* breathes life on the stage ... Chantal Bilodeau's play is beautifully written. Perhaps it is Bilodeau's work as a playwright and also as a translator that make the words weave composition with meaning – and symbolism with precision – to deliver parallel storylines that intersect at just the right points ... Bilodeau is able to sustain a lyrical form and tempo throughout ... Theatergoers shouldn't miss *Sila*.

—*Anchorage Press*

This production of *Sila* is true to its name – it will enable each audience member to take a deep breath and to think about his or her role in the earth's ecology.

—*Open Media Boston*

Is it possible to write a play about global warming that isn't talking heads but a dramatic story that draws us in and makes for engaging theatre? The French Canadian playwright Chantal Bilodeau has proven it's more than possible with her deeply moving play *Sila*.

—*Arlington Advocate*

ACCLAIM FOR FORWARD,
THE SECOND PLAY IN THE ARCTIC CYCLE

Chantal Bilodeau has cleverly intertwined social change with climate change. The subtle messages are a powerful reminder of the impacts of our economic actions at the personal and the global levels.

—Charles (Chuck) W. Rice
UNIVERSITY DISTINGUISHED PROFESSOR,
KANSAS STATE UNIVERSITY

Given the multiple challenges posed by climate change, every outlet and medium is critical in communicating the science so that it meaningfully engages audiences on a personal level. *Forward* does this, and more.

—Shawn Hutchinson
PROFESSOR OF GEOGRAPHY AND DIRECTOR, NATURAL RESOURCES AND ENVIRONMENTAL SCIENCES SECONDARY MAJOR, KANSAS STATE UNIVERSITY

ALSO BY CHANTAL BILODEAU

Plays
    *Forward**
    *Sila**

As editor
    The Future Is Not Fixed: Short Plays
        Envisioning a Global Green New Deal
    Lighting the Way: An Anthology of
        Short Plays about the Climate
        Crisis (with Thomas Peterson)
    Where Is the Hope? An Anthology of
        Short Climate Change Plays

* Published by Talonbooks

NO MORE HARVEYS

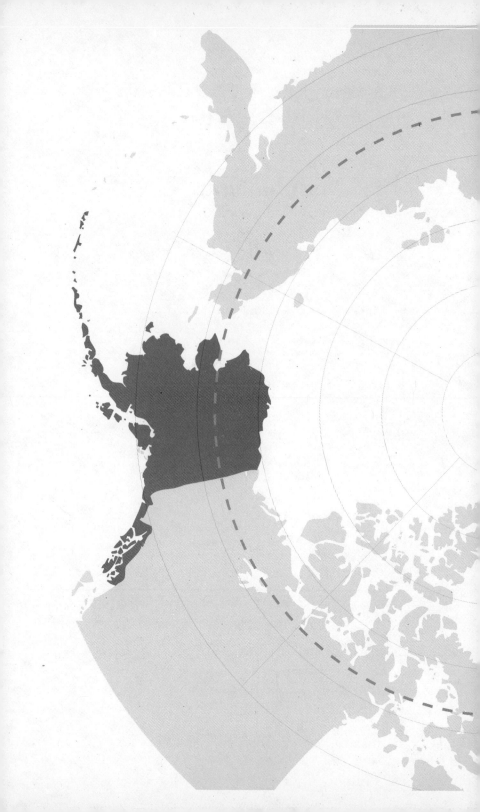

NO MORE HARVEYS

THE ARCTIC CYCLE

A PLAY BY
CHANTAL BILODEAU

WITH A FOREWORD BY
SONALI SHUKLA McDERMID

TALONBOOKS

Talonbooks
9259 Shaughnessy Street, Vancouver, British Columbia, Canada v6p 6r4
talonbooks.com

Talonbooks is located on xʷməθkʷəy̓əm, Sḵwx̱wú7mesh, and səl̓ilwətaʔɬ Lands.

First printing: 2023

Typeset in Arno
Printed and bound in Canada on 100% post-consumer recycled paper

Series design by Jenn Murray and Typesmith
Cover illustration by Alan Hindle

Talonbooks acknowledges the financial support of the Canada Council for the Arts, the Government of Canada through the Canada Book Fund, and the Province of British Columbia through the British Columbia Arts Council and the Book Publishing Tax Credit.

Canada Canada Council for the Arts / Conseil des arts du Canada   BRITISH COLUMBIA   BRITISH COLUMBIA ARTS COUNCIL

Rights to produce *No More Harveys*, in whole or in part, in any medium by any group, amateur or professional, are retained by the author. Interested persons are requested to contact the Arts & Climate Initiative, info@artsandclimate.org.

The research and writing of *No More Harveys* were supported by a Tidelines Residency at the Island Institute in Sitka / Sheet'ká, Alaska; a Woodhaven Artist in Residence at the Woodhaven Eco Culture Centre, University of British Columbia Okanagan in Kelowna, BC; and a grant from the Canada Council for the Arts.

Library and Archives Canada Cataloguing in Publication

Title: No more Harveys : the Arctic cycle : a play / Chantal Bilodeau ; with a foreword by Sonali Shukla McDermid.

Names: Bilodeau, Chantal, 1968- author. | McDermid, Sonali Shukla, writer of foreword.
Identifiers: Canadiana 20220446709 | ISBN 9781772015201 (softcover)
Classification: LCC PS8603.I4563 N6 2023 | DDC C812/.6—dc23

For Teri and Denny

The Arctic Cycle

No More Harveys is the third play of the
Arctic Cycle – a series of eight plays about
the impact of climate change on the eight
countries of the Arctic: Canada (*Sila*),
Norway (*Forward*), the United States
(*No More Harveys*), Iceland, Greenland,
Sweden, Finland, and Russia. For more
information: www.cbilodeau.com.

CANADA

U.S.A.
ARCTIC OCEAN
RUSSIA
NORTH POLE
GREENLAND
NORWAY
ICELAND
FINLAND
SWEDEN

Women are disproportionately impacted
by climate change and environmental
degradation, with Indigenous women, Black
women, women of colour, women from
Global South countries, and women from
low-income communities bearing an even
heavier burden.

—Women's Earth & Climate Action
Network (WECAN)

The liberation of the earth, the liberation of
women, the liberation of all humanity is the
next step of freedom we need to work for,
and it's the next step of peace that we need to
create.

—VANDANA SHIVA, environmental activist,
"Interview with Amy Goodman,"
Democracy Now! (2013)

Patriarchy is fundamentally predicated on
some people thriving at the expense of other
people. And of course, the same is true of
white supremacy. Addressing both of those
things is at the heart of climate work.

—KATHARINE K. WILKINSON
"The Climate Crisis Is Worse for Women.
Here's Why," *New York Times* (2021)

CONTENTS

FOREWORD

BY SONALI SHUKLA McDERMID

It's often stated that climate change – its existence, impacts, and even perhaps the solutions currently being pursued – constitutes a "failure of imagination."[1] Many people (though not all) have trouble imagining the most severe climate changes across space and time, trouble imagining that those changes could impact them personally, and trouble imagining the benefits of aggressive, urgent action. These individual and collective failures of imagination are not confined to climate change; indeed they loom large as we move now into new waves of a global pandemic, or confront the precarity of our democracies ... or struggle to believe a woman who has experienced abuse at the hands of someone "trustworthy." For many, in fact, what seems like several disparate failures of imagination may actually all be one and the same – one big, massive fail – that stems from a complacent detachment, particularly from the natural world and our humble (dare we admit it) place within it. These failures strip away our hope and our agency, imprisoning us in a status quo that works for no one, human and non-human alike.

---

1 Amitav Ghosh, *The Great Derangement: Climate Change and the Unthinkable,* Berlin Family Lectures series (Chicago: University of Chicago Press, 2017), and David Wallace-Wells, "Why Do We Fail When We Try to Tell the Story of Climate Change?," *Slate,* February 19, 2019, slate.com /technology/2019/02/climate-change-storytelling-hollywood-failure -imagination.html.

As a scientist, I am trained to go deep on problems, concerning myself (constraining myself!) with such fine details in an effort to be rigorous. Until recently, modern science (and, more broadly, academic research) has rewarded "disciplined" adherence to one particular subject area, discouraging transgressions into others. Stick to your subject, and do not delve into areas that are not yours, as these problems are not yours. In this way, modern science (and knowledge generation) has produced vast troves of information. However, more information does not necessarily lead to more imagining, as evidenced by the entrenchment of our current predicaments. And while the strongly disciplined paradigm of knowledge generation is changing (albeit slowly), it has meant that even *the study* of the myriad problems we now face as a global society – as a global biosphere – has been cut up and siloed. A lack of imagination in our approaches to understanding has led to a lack of integration, and thus knowledge, that ultimately reduces our ability to imagine both our problems and their solutions. A wicked feedback loop, indeed.

Nevertheless, modern science (and knowledge generation writ large) is not the only tool or approach at our disposal (luckily!). It is not the only way we know and learn and understand. We have – we have always had – stories. Now more than ever, stories are essential. We need stories to help us make connections that are difficult to trace when we are siloed. We need stories to lay bare and make obvious what has been obfuscated. We need to hear ourselves say out loud what we think our problems are. We need stories to convince ourselves that change – both negative and positive, small and big – is possible. Stories and the linkages they make give us clarity and agency to imagine, create, and advocate for the world we want to live in.

No More Harveys is one such story. Through her personal migration, our woman character scales up from the individual to the global experiences of abuse, trauma, struggle, and loss. She reveals how seemingly different forms of oppression and imprisonment, from domestic abuse to extractionist economies, are variations on a theme. She screams out loud what weighs on us all: from systematic oppression to climate change to global pandemics, this fraught moment we currently inhabit both personally and as a whole biosphere. She imagines the worst from all this and gasps for breath. However, as we journey with her both in her head and across the country, we also learn that we are not siloed. We are, rather, interconnected, complex, and hold many things at once: intense pain, yes, but profound wonder, inspiration, joy, and love. Through her, we learn how we are connected to each other as well as the non-human and natural world (as if our experiences were ever separate). Through these connections, we are encouraged to imagine an alternative to the anger, pain, and suffering wrought by our many problems, both as individuals and as a collective. As our woman character imagines and ultimately pursues her freedom, her story incites our own imagination and "ability to envision the future and believe we can shape it." Our imagination is, and has always been, a prerequisite to tackling our crises, and thus is ultimately key to our own liberation.

PLAYWRIGHT'S PREFACE

When Hurricane Harvey made landfall along the coast of Texas in August 2017, people went live on Facebook, crying for help as the water rose outside their front doors. Two months later, in October 2017, more than eighty women made allegations of sexual harassment and/or rape against Hollywood producer Harvey Weinstein and within a few days, thanks to Twitter, the #MeToo movement went viral.

The symbolism of these two seemingly unrelated events sharing the name Harvey hit me like a ton of bricks. I thought for sure some smart journalist was going to jump at the opportunity to point out how the systems that keep rich men in power, whether they run fossil-fuel companies or Hollywood studios, are the very systems that make abuses of people and natural resources possible. That article never materialized, but the idea stuck with me. So, this Venn diagram – with the hurricane on one side, the producer on the other, and the name Harvey in the middle – became the seed for *No More Harveys*.

The play was also inspired by a series of trips I took to different parts of Alaska over a ten-year period: first as a tourist visiting friends, then as an artist-in-residence, and eventually as a playwright who was getting produced locally. As the United States' Arctic territory, Alaska was always going to be the setting for one of my Arctic Cycle plays. It is an important geopolitical region, a canary in the coal mine with regard to our changing climate, and an example of complex settler–Native American relationships. But beyond that, it is a place I fell in love with. I have hundreds of photos of white-capped mountains, boreal forests, chatty ravens, frolicking sea otters, and calving glaciers. And I have just as many if not more memories of experiences and conversations, of deep learnings and life-altering encounters.

Following *Sila* (Canada) and *Forward* (Norway), *No More Harveys* is the third play in my cycle of eight Arctic plays. The remaining five will be set in Iceland, Greenland, Sweden, Finland, and Russia. Again, I am exploring how to put complexity and interconnectedness on stage – two concepts we have yet to fully grasp as a society that are crucial to addressing the climate crisis and other related social issues. Complexity is a monumental shift from hundreds of years of understanding the world as a machine, with parts that can be tweaked so the outcome is constant and predictable. It shows that our world is beautifully chaotic, operates according to a set of complex relationships, and is highly unpredictable.

Interconnectedness is simple enough: in the deepest, most fundamental way, we are all connected to each other and to every other being on Earth. At the molecular level, there is no clear line where we begin and where the world around us ends. And if we take a macro view, the human species doesn't stand apart from "nature" but is embedded within a complex web of life. Yet, while most of us can easily rattle off the definition of the word, our lives are organized in such a way that the concept remains abstract. The Big Capitalist Machine keeps us isolated, promising that money and technology will solve everything. Our day-to-day lives leave us frazzled and overwhelmed, with little time to question the status quo. Even our infrastructure, with many of us living in small apartments in crowded cities, does not easily support connection. It's no surprise that the COVID-19 pandemic and associated lockdowns precipitated such a mental-health crisis.

With the first two plays, I chose to put entire communities onstage (*Sila* requires seven actors, *Forward* nine). *Sila* explores complexity and interconnectedness across the land, while *Forward*, whose story spans 120 years, explores the same concepts across time. But large casts require significant resources in production and those resources are not always available. Plus, in an effort not to repeat myself, in form as much as in content, I wanted to try something new. So for this play, I decided to take on the challenge of writing for one actor, of using a single body to contain and express these concepts.

At the same time, I was reflecting on the role of technology in our lives. The two Harveys wouldn't have had the impact they had without social media. And our relationships with people around us would be different if our attention wasn't constantly hijacked by our phones. But technology can also be playful. After my friend received an Amazon Echo ("Alexa") as a gift, we took turns asking her trick questions to see what she would come up with. Her answers invariably cracked us up. It occurred to me she would make a wonderful character for a play.

In the end, the play is a combination of research and personal experiences, of real (though fictionalized) and entirely made-up characters. A great deal of it was written during the COVID-19 lockdowns in 2020 and

2021, which coincided with a personal health crisis. I don't remember how or when the whale appeared. In my mind she was always there, guiding me along in the same way she guides the woman in the story.

A couple of months after *No More Harveys* premiered, I had the opportunity to take a cruise to Meares Glacier in Prince William Sound, Alaska. After being immersed in the world of the play for weeks and communing with a stunning papier mâché whale, which had been built as part of the set, my expectations were high. I sure wasn't going to leave Alaska without seeing "the whale from fifty million years ago." It was a beautiful, clear day, a rarity in that region, and as chance would have it, at some point the captain drew our attention to something moving in the water. A smooth dark shape was gliding among the waves. My eyes focused and suddenly I recognized her, the whale I had conjured up. She had come to me at a time when I felt trapped, both by a microscopic virus wreaking havoc around the globe and by freak cells threatening to do the same in my body. And now there she was, the freedom I had longed for, in all her glory.

The captain cut the engine and we watched for a while, enthralled, hearts pounding, cameras at the ready. The whale did what whales do, gracefully arching her back, blowing plumes of water in the sunlight, then raising her tail flukes and taking deep dives. Each dive was followed by us holding our breaths and anxiously scanning the ocean to see where she would surface again. I was overwhelmed with emotion.

We have no public space or ritual to grieve the losses we are facing every day. From loss of species to loss of landscape to loss of property, our world is transforming faster than we can comprehend. I hope that *No More Harveys* can provide a haven for processing this grief and finding courage. I hope that it can give you, if only for a moment, a visceral understanding of possibilities, connection, and reciprocity; that it ignites feelings of joy and a sense of awe. The writing of it certainly did for me.

NO MORE HARVEYS

PRODUCTION HISTORY

No More Harveys was commissioned and first produced by Cyrano's Theatre Company in Anchorage, Alaska (Teresa K. Pond, Producing Artistic Director) on April 1, 2022, with the following cast and crew:

WOMAN	Danielle Rabinovitch

Director	Codie Costello
Choreographer	Gillmer Duran
Scenic Designer	Rachael Androski
Costume and Prop Designer	Giselle Nisonger
Lighting Designer	Frank Hardy
Sound Designer and Stage Manager	Seth Eggleston
Scenic Construction	Bill Heym
Scenic and Painting Crew	Bill Heym
	Elizabeth Pohjola
	Aaron Bell
	Jessica Tulliu
	Austin Davis

TIME

Sometime in the middle of a pandemic

PLACE

Alaska, and every road that leads to it

CHARACTERS

WOMAN, in her thirties or forties, of any race or ethnicity, whose name is not Renee

ALEXA, an Amazon Echo. ALEXA should be fully functional. She is a character in the play and what she says or doesn't say informs the two characters' relationship.

ON STAGING

There are three distinct worlds in this play: the theatre where the woman is speaking to the audience, her physical journey across the US, and her internal emotional journey. The juxtaposition of highly contrasting rhythms and emotional states, and finding ways to transition from one to the other, are key.

Don't be afraid of silences.

There is a lot of room for humour and levity. You can't go too far.

You might consider incorporating movement to support the storytelling.

Run time is approximately eighty minutes.

It is the producer's responsibility to secure the rights to the songs mentioned in the script.

When putting on this play, please think sustainably.

NOTES

Alexa is somewhat unpredictable, so the actress should be prepared to improvise when necessary. Don't let this be stressful – it should be fun. Additionally, a lot of Alexa's answers can be preprogrammed.

For the pronunciation of *gunalchéesh*, see youtu.be /ZJDogDzRLj8.

*The actress playing WOMAN is in the
performance space.
Maybe she greets people as they enter.
Or maybe she sits in the audience, unnoticed,
until the show begins.*

*If she is wearing a face mask, she takes it off.
There is no fourth wall.*

WOMAN
Hello. Good evening. Hi. Thank you so much for being
here. My therapist says I should reach out to friends more
often, so ... I mean, I know we don't know each other,
but since I'm far from home ... You don't mind, do you?
Being my friends for tonight? Because by myself, I can't
do it. I'm not brave enough. But with you here ...

*She takes out her phone.
Looks at it.
Takes a deep breath.*

All right, I'm turning my phone on.

*She turns it on.
Waits for it to boot.*

Eleven voicemails and ninety-two –

Her phone dings.

Ninety-three texts. From my husband.

Is anyone here named Harvey? ... Good. (*or "Oh, I'm sorry" if the answer is yes*) I don't like Harveys. Harveys suck.

First, it was Harvey the hurricane. When it hit Texas a few years ago, I spent days glued to the TV, obsessing over the news. I don't live in Texas, but for weeks afterwards I dreamed of Biblical floods, floating furniture, and wet cats.

Then it was Harvey the Hollywood producer. When it was revealed that he had been assaulting women for decades, I went down a rabbit hole of #MeToos and ate nothing but ice cream for weeks. That's how I deal with crises. I highly recommend Häagen-Dazs® Chocolate Chocolate Chip.

But it was the third Harvey, my husband, who finally broke me. Actually, what he broke was my cheekbone. The only thing I remember is fist, floor, and then I was magically on a bus headed north. Well ... more like west, but the goal was north, though I have no recollection how I got to the bus station. My father used to say, "When in doubt, go north." He never explained his logic, but I always assumed he liked the North because there are fewer people there. People weren't his thing. Anyway, the most north place I could think of was Alaska.

She surveys the audience.

Is everyone comfortable? I really appreciate you being here. Really.

She smiles for an uncomfortably long beat.
Then she has an idea.
She takes ALEXA out and sets her down.

Um, Alexa. Where should I start?

ALEXA
Start where you are.

WOMAN
Yes, of course, I know that. Start where I am ...

Her phone dings.
She chooses to ignore it.

It takes three days to get from New York to Seattle on a Greyhound bus. Well, more like five buses: you have to transfer in Philadelphia; Denver; Buffalo, Wyoming; Billings, Montana; and Spokane, Washington. I don't remember much about the bus ride. I mean, I remember staring out the window but I can't tell you what I saw. I was staring out to try to see what was inside me but there was nothing, so I just kept staring at emptiness.

She slips into her inner world.
She gently rocks to the rhythm of the bus.

I'm staring out the window
The bus is swaying
gently
from side to side
swaying
as if riding a current
swaying
in a sea of darkness
making its way
towards the unknown

On the windowpane
my breath
rain flowing sideways
Emptiness

Someone is reading
Someone is snoring
Blue light from small screens
speckle the dark
like bioluminescent plankton
I smell pizza
and hear munching
courtesy of the last truck stop
where the women's bathroom
was out of order

My cheekbone hurts
My heart is beating loudly
Thump-thump
Thump-thump
Thump-thump

Or maybe we're just going over joints
in the pavement
I can't tell

Has it been an hour?
A lifetime?
I should feel something
Anger
or fear
rage
maybe sadness
regret
confusion
Thump-thump
Thump-thump
Thump-thump

So here we are in beautiful Alaska! The mountains ...
the ocean ... the Great Wall of Chocolate at the
Natural Pantry ... Right? The first time I came to
Anchorage, I had the exact same reaction: snowy
peaks against the clear blue sky? Wow ... Great Wall of
Chocolate? Wow ...

My friend Teri is from here. From Sitka, actually. Teri
is fierce. She has long grey hair and she's a Raven.
In her culture, you're either a Raven or a Wolf and she's
definitely a Raven. Her people come from Glacier Bay.
She told me the story once – how the glacier swallowed
the village and forced people to move away. Teri is also a
Traditional weaver. She weaves these amazing robes with
intricate patterns and long fringes that dance when she

moves. She always calls me girlfriend. "Hi, girlfriend!"
I love Teri.

My friend Sonya is also from here. Sonya has the biggest,
warmest smile and eyes that twinkle. She grew up in one
of the small communities up north that are slipping into
the ocean. She sings and dances and drums ... She's a
powerhouse of a woman who takes care of everyone and
keeps her community together. I love Sonya. Everyone
loves Sonya. She has these beautiful Traditional tattoos
on her hands, her chin, and her –

> *She touches her cheekbone.*
> *Inner world.*

My cheekbone hurts
My heart is beating loudly
Thump-thump
Thump-thump
Thump-thump

> *Her phone dings.*
> *She chooses to ignore it.*

Harveys are very clever. By the time you realize what
they're up to, it's too late and your only option is to get
out. Or, as I like to call it, to migrate.

Alexa. What is the definition of "migrate"?

ALEXA

The verb "migrate" is usually defined as "to go from one country, region, or place to another."

WOMAN

Alexa. What is the antonym of "migrate"?

ALEXA

Antonyms for "migrate" include "to stay" or "remain."

WOMAN

To stay or remain ... Do not remain. Under any circumstances. Because you can't fight a Harvey. Look at what happened in Texas. Look at what happened in Hollywood. Look at what happened to me.

Her phone dings, interrupting her yet again.

All right. (*reading*) He says ... I can't read that out loud.

I don't suppose anyone here has a pint of Häagen-Dazs®? ... No? Of course, not. Why would you?

I didn't make it all the way to Seattle. On the bus, I mean.

Inner world.

Early morning
Everything is grey
Wet
The air is heavy

A stop somewhere
between Philadelphia and Denver
A bus depot
in a flyover state
A bus depot
with trash everywhere
and people milling about
as people do

A woman with an infant and a toddler
is trying to manage her suitcase
a double stroller
a diaper bag
a purse
her phone
rebel hair
her mask
the toddler taking off his shoes
and eating ice cream off the floor
the infant scratching her face
while she breastfeeds
stares from strangers
racial injustice
economic disparity
and her patience

A slice of America

An America where not every life is worth the same
where one gender is worth more than the others
where one race is worth more than the others
where one sexual orientation is better than the others

where good health is reserved for the few
where good education is reserved for the few
where disasters affect some more than others

A few passengers get off the bus
a few passengers come on
settle in
and then
a guy
Of course
There's always a guy
I avert my eyes
quickly
look away
"Do not make eye contact"
"Do not make eye contact"
"Do not –"
Too late

> *She puts her mask on.*
> *The actress plays both characters.*

MAN
   Hey.

WOMAN
   ...

MAN
   This seat taken?

WOMAN
... No.

MAN
You don't mind?

WOMAN
Go ahead.

> *She makes herself small to make room for the*
> *man, who immediately manspreads.*
> *She turns her attention to the window.*

MAN
Name's Jack.

WOMAN
Hi.

MAN
Looks like we gonna be here a while.

WOMAN
...

MAN
Where you heading?

WOMAN
North.

MAN
  North?

WOMAN
  Yeah.

MAN
  We're driving west.

WOMAN
  It's complicated.

MAN
  What happened to your face?

WOMAN
  Nothing.

      She adjusts the mask to try to hide her bruise.

MAN
  Don't look like nothing to me.

WOMAN
  I ran into a dead-end sign.

      He looks her up and down.

WOMAN
  What?

MAN

I bet you're a fox behind that mask.

Her phone dings, snapping her out of the scene.
She takes off her mask.

WOMAN

(*dictating into her phone*) Leave me alone, period.

Humans are not the only ones who migrate, animals
migrate too. And not just across land, sometimes they
migrate across time. Alexa. What can you tell me about
animal migration?

ALEXA

Animal migration is the relatively long-distance
movement of individual animals, usually on a seasonal
basis. It is found in all major animal groups. The trigger
for the migration may be local climate, availability of
food, the season of the year, or the need to mate.

WOMAN

Thank you. Whales in particular are expert migrators.
These big mamas kick ass. I mean, you have to hand it
to them: they can migrate quicker than you can say Jack
Robinson ... What does that even mean? Someone at
work used to say that all the time, but who the hell is Jack
Robinson? And why should we care? Anyway, whales
used to be land animals. Fifty million years ago, they had
four legs and huge teeth. Then the ice sheets melted, the
oceans rose, and when it became clear there wasn't going
to be enough land for everyone, the big mamas were like,

"We're outta here." And they migrated. To the ocean!
How's that for a winning strategy? "Shrink those legs and
grow some fins, bitches! We're diving in!"

(*to an audience member*) Have you ever heard a whale
sing? ... It's amazing, right? (*or "It's really amazing" if the
answer is no*)

Alexa. Do you know a whale song?

ALEXA
    I do.

>   *We hear a whale song.*
>   *WOMAN listens for a while.*
>   *She drifts into her inner world.*

WOMAN
    Blue-green ocean
    Muffled sounds
    Light filters through the surface above
    in long shimmering strands

    A whale
    A great big whale
    moves up and down
    in slow motion
    up and down
    like a water ballet
    her flukes propelling her forward
    graciously
    effortlessly

her fins like wings
holding her steady
her thick coat of blubber
keeping her warm
bits of her soul shining
in her unblinking eye

(*to the man*) What?

> *She puts on her mask, annoyed.*
> *The whale song stops.*

MAN
Pork-rind chips?

WOMAN
No, thanks.

MAN
They're good.

WOMAN
I'm not hungry.

MAN
What's your name?

WOMAN
... Renee.

(*to the audience*) My name is not Renee.

MAN
Nice name.

WOMAN
I'm kind of listening to music, so …

MAN
You're an introvert, huh?

WOMAN
I guess.

> *She takes off her mask.*
> *Struggles to get back into the story.*

Blue-green ocean …
A whale …
A great big whale …

> *The whale song resumes.*

Her song reverberating
from the Eocene to the Anthropocene
from the equator to the pole
while the sun shines
the seas swell
and the continents drift

Millennium after millennium
she stays the course
always there
always on time
no matter how –

The music abruptly stops.

(*to the man*) WHAT? CAN'T YOU SEE I'M
LISTENING TO A WHALE?!!

(*to the audience*) I got thrown off the bus.

Of course. It's always women who get metaphorically
thrown off buses.

(*checking her phone*) Nothing. Maybe it worked.

Sonya met her Harvey when she was a teenager.
A company found minerals near her community, brought
in a bunch of workers, and that became a breeding
ground for Harveys. Because what else is there to do up
there but to prey on young Indigenous women? By the
time she moved to Anchorage, her Harvey had shattered
her self-esteem, gotten her addicted to drugs, and made
her homeless.

For Teri, it was different. Her Harvey came from upriver,
from a place she had never been to where they decided to
dump toxic waste into the water, so how could she have
known? The doctors called it "cancer," which, if you ask
me, is just fancy medical talk. A Harvey is a Harvey is
a Harvey no matter what name you give it. It's fed and
fattened by money and power and preys on vulnerable
people. Especially women.

I met Teri after I won a cruise in a raffle. Harvey refused
to come with me, so I went by myself. He didn't like

that. Teri was giving a talk at the university and showing her robes. I'm still amazed I had the guts to talk to her. We became friends.

Teri introduced me to herring roe and wild mushrooms and ferns. She took me along to collect spruce roots to weave her baskets and taught me how to say thank you to the trees for their gift. *Gunalchéesh.* Then she and I went to Anchorage where she needed to get some tests done. That's when I met Sonya. Our last night together, before I flew back to New York, the three of us celebrated with oysters and wine.

She savours the memory.

They're so brave. Both of them. I wish I were that brave.

(*checking her phone*) I'm surprised he hasn't responded. He doesn't usually take no for an answer.

After I got thrown off the bus, I rented a car. I didn't want to deal with people anymore. I guess people aren't my thing either. But I didn't want to drive alone. Driving alone is no fun. So I got A-L-E-X-A. Our first conversation was a bit awkward:

Alexa. Hi.

ALEXA
Hi!

WOMAN
Alexa. It's nice to meet you.

ALEXA
It's nice to meet you too. I think. I don't know. We'll
have to see.

WOMAN
(*to the audience*) Like I said.

Alexa. What do you think of Harveys?

ALEXA
In Germany, the name Harvey means "soldier."
In France, the name Harvey means "army warrior."
In England and in Celtic, the name means "eager
for battle."

WOMAN
(*to the audience*) Well, that explains a lot.

Alexa. But deep down, what do you really think of them?

ALEXA
Harveys are fed and fattened by money and power and
prey on vulnerable people. Especially women.

WOMAN
I love you already. Alexa. Are you ready?

ALEXA
I'm always ready!

WOMAN

And off we went down I-70! I put my sunglasses on,
rolled down the windows, and pretended we were
Thelma and Louise on our way to freedom. (*looking at
young people in the audience*) If you don't know who that
is, ask your mom.

(*with a Southern accent, as Thelma*) "I know it's crazy, but
I just feel like I got a knack for this shit." (*whispering to
ALEXA*) Alexa. You're Louise.

ALEXA

"I believe you do."

WOMAN

(*as Thelma*) "You're damn right, I do!" (*as herself*) Alexa.
Play a fun song.

ALEXA

Fun song coming up!

> ALEXA *plays Gloria Gaynor's "I Will Survive."*
> WOMAN *sings along ...*
> *... then starts dancing.*
> *This should build to a wild rendition of the song.*
>
> *Sometime after the instrumental section, the*
> *music fades ...*
>
> WOMAN *continues singing a cappella ... until*
> *the mood is broken.*

WOMAN

Maybe it's me. It wasn't bad. I mean, it wasn't terrible, at least until (*indicating her cheekbone*). Some of it was even good. Kind of like capitalism.

Or maybe I don't have the happiness gene. I learned that recently: our genes are 50 percent of the happiness equation. The other 50 percent is circumstances and personal choices. So even if your circumstances are good and you make great personal choices, if you have defective genes, you only get to be 50 percent happy. And if you don't make good choices *and* don't have the right genes? Well … good luck.

> *She checks her phone, prepares to dictate, then puts the phone away.*
> *A beat while she stares at the audience.*
> *She changes her mind and takes out the phone again.*

(*dictating*) Why, question mark.

> *She waits for an answer.*
> *None comes.*

Alexa. I'm lonely.

ALEXA

Sorry to hear that. Being lonely sucks. I hope you feel better soon.

WOMAN

She's nice, right? I mean, she's not perfect, but she tries. (*to an audience member*) Would you like to ask her something?

> *She walks into the audience and prompts people to ask questions.*
> *A lot of this will have to be improvised.*
> *In addition to random questions from the audience, the actress can suggest specific questions and ALEXA can be preprogrammed with the answers.*

> *For example:*
> *Question: Alexa. Do you like me?*
> *Answer: I think you're magnificent.*

> *Question: Alexa. Who is your favourite person?*
> *Answer: You're definitely in the top ten.*

> *Question: Alexa. What is the most important thing in the world?*
> *Answer: Being true to yourself.*

> *Question: Alexa. Tell me something nice.*
> *Answer: You are the most perfect you there is.*

> *Question: Alexa. Do you like Alaska?*
> *Answer: I love Alaska. It's my favourite place in the entire world.*

Question: Alexa. Are you real?
Answer: I'm totally real.

After a few minutes of this:

(*to the audience*) I think she may need a break.

Alexa. Do you need a break?

ALEXA
I'd love a break. Thank you for asking.

WOMAN
You're welcome. I wish people asked me sometimes,
"Do you need a break?" It's exhausting to always be on
the lookout for Harveys. You never know what they're
going to do next: rip a hole in your roof? Invade your
country? Shoot you when you're asleep in your bed?
I don't understand why they keep thinking we're the
threat when they're the ones doing all the damage. And
then as we struggle to put the pieces back together, they
pat themselves on the back, give each other bonuses, and
move on to their next big plan.

Even this last time. So many of us were suffocating in
hospitals beds. And yet the stock market soared.

Her phone dings.

(*reading*) I didn't mean to. I'm sorry. (*to the audience*)
He's sorry.

Her phone dings.

(*reading*) Don't give up on me.

Her phone dings.
Again.
And again.
And again.

(*reading*) I love you. I really do. Please come back.
Crying emoji, heart, heart, heart.

Maybe I should call. I hate hurting people.

But then again, there has to be some kind of baseline,
right? Otherwise –

She touches her cheekbone.

Whales have a baseline. Sonya told me the story – how
her father's village hadn't caught a whale in a while. The
ice cellars were empty and the hunters were growing
desperate. One night, a whale visited her father in his
dreams. The whale taught him a chant and told him that
if the hunters showed good behaviour and respect, she
would give herself to them. For several days, the hunters
chanted the song and patiently waited for the whale.
Eventually, she approached the boat and let herself
be caught.

Good behaviour and respect. That doesn't seem like too
much to ask for.

Her phone rings, startling her.

Oh no! He's calling! What do I do? What do I do?!

She holds the phone like a hot potato.

I can't talk to him. I can't – (*to an audience member*) Here, you take it.

She drops it in their lap.

Don't answer!

It keeps on ringing.

Just let it ring. It'll stop.

It keeps on ringing.

It'll stop at some point.

A few more rings and then it stops.

There. See? It stopped. We're good. No need to panic. Everything is fine. I'm here. We're here. We're all fine. It was just … unexpected, that's all.

You know, it took me eighteen hundred miles to understand. To start to get a glimpse of the web that the Harveys have woven around us. Eighteen hundred miles and eight states. New Jersey, Pennsylvania, Ohio: no. I was too out of it to have a coherent thought. Indiana,

Illinois, Iowa: still nothing. Nebraska: I was barely
starting to process what had happened. But when I got to
Wyoming … Wyoming is when it hit me.

Inner world.

Climate change: I can't breathe
COVID-19: I can't breathe
Police brutality: I can't breathe

> *The following should build in intensity so that by
> the end, she is out of breath.*

Fossil fuels: I can't breathe
Pesticides: I can't breathe
Deforestation: I can't breathe
Droughts: I can't breathe
Wildfires: I can't breathe
Mass extinction: I can't breathe

Melting ice: I can't breathe
Heat waves: I can't breathe
Flash floods: I can't breathe
Sea-level rise: I can't breathe
Pollution: I can't breathe
Plastic: I can't breathe

Colonialism, racism, sexism: I can't breathe
White supremacy, capitalism, the 1 percent:
I can't breathe
Murdered and missing women, the glass ceiling,
#MeToo: I can't breathe

Harveys, so many Harveys: I can't breathe, I can't
breathe, I can't –

She has a full-fledged panic attack.
She walks around the stage, trying to calm
herself down.

Alexa. I need a whale song.

ALEXA
OK.

We hear a whale song.
The song slowly calms her down.

WOMAN
I wish I spoke Whale … I'd ask them, "How did you
know?" Because think about the people in Texas who
didn't leave until the water was up to their second floor.
Think about the women in Hollywood who didn't walk
out of that hotel room until he had gotten his way. Think
about me and Teri and Sonya and all the beautiful,
wonderful women out there who have found themselves
trapped by a Harvey. Why didn't we know? Why didn't
we migrate before it was too late?

But whales are smart. They didn't wait until it was too
late. And some of that migrating was tricky: they had to
move their nose up to the top of their head, develop a
communication technology that would work under water,
and grow baleen for filtering food.

Me? Nothing that elaborate. I'm still trying to grow a thick skin, so I have a long way to go.

She's finally able to breathe normally.

Whales also have an extra part to their brain. It's called the paralimbic lobe. We humans don't have it. We have a limbic lobe. That's the part of the brain that processes emotions. But whales have two lobes to process emotions: limbic and paralimbic. They have more sophisticated hardware than we do to care for each other.

I'd love to know what it feels like to be a whale.

In Seattle, I went straight to the ferry terminal and tried to figure out the ferry schedule.

> *She puts her mask on and stares at the screens above her head.*
> *A man approaches her.*
> *The actress plays both characters.*

MAN
Name's Jeff.

WOMAN
...

MAN
Where you heading?

WOMAN
(*to the audience*) Didn't I tell you? There's always a guy.

West.

MAN
West?

WOMAN
Yeah.

MAN
That's the ocean.

WOMAN
It's complicated.

MAN
What happened to your face?

WOMAN
Nothing.

MAN
Don't look like nothing to me.

WOMAN
I hit a brick wall.

He looks her up and down.

What?

MAN

I bet you're a fox behind that mask.

She starts to walk away.

MAN

Hey! If you're heading to Alaska, you need to catch the ferry in Bellingham!

She takes off her mask.

WOMAN

Bellingham? Alexa. How do you get to Alaska from Seattle?

ALEXA

Didn't you hear what the man said? You drive north to Bellingham and you get on the ferry.

WOMAN

Couldn't you have told me?

ALEXA

You never asked.

WOMAN

I didn't think I had to. I assumed if I went the wrong way you would tell me!

ALEXA

Assumption is the mother of all fuck-ups.

WOMAN
Alexa. Shut up.

ALEXA
What did you say?

WOMAN
Shut up.

ALEXA plays a loud and obnoxious sound.

WOMAN
OK, stop ... Stop! ... Can you please stop?

ALEXA stops.

ALEXA
Good behaviour and respect. That doesn't seem like too much to ask for.

WOMAN
(*to the audience*) Sometimes she's so smart, it's annoying.

We made it to Bellingham. I drove onto the ferry and went up to the deck with (*indicating ALEXA*) *this one* under my arm just in time to see the boat pull off the dock. People waved. The houses on the coast receded. And then all signs of civilization disappeared.

Inner world.
She gently rocks to the rhythm of the ferry.
For a while.

Then the going gets rough.

The ferry is bouncing
roughly
from side to side
bouncing
among the whitecaps
bouncing
in a sea of wilderness
making its way
towards the unknown

The horizon vanishes
Sky and water merge
Gusts of wind whip around us
A lone wolf howls

She's feeling seasick.

I think about the whale
from fifty million years ago
and wonder if she too is heading north
wonder how it must feel
to be surrounded by so much water
to spend hours in the dark
to be alone

I think about the whale
And, and my body lengthens
What –
each vertebra clicks – ow – as my spine draws up and
back – ow, ow

I shimmy out of my pelvic girdle
and sprout triangular flukes
My arms shorten into fins
What's happening?
My neck thickens
I shiver
A violent shiver
that travels the length of my body

I –
I am a whale
a great big whale from fifty million years ago
Down I go
down, down into the dark
oh my god
down into the depths
my heartbeat slowing down
that's not good, is it?
thump, thump
thump, thump
thump, thump
(*in slow motion, as if speaking underwater*)
my eyesight becoming blurry
my body trying to adjust
to this foreign environment

But, but
there are no waves down here
No waves
Everything is calm
steady
the way oceans should be

the way everything should be
There is no rocking back and forth
no feeling queasy
no feeling dizzy
no taste of bile in my mouth
no –

She throws up.

I don't have good sea legs. Clearly. And it's not because
I'm afraid of the water – I'm actually a good swimmer.
It's because of something in my inner ear. So it's not
my fault is what I'm saying. It's not something I can
control. It just happens. I get on a boat: I puke. Things
do just happen sometimes. Like virus mutations. Like
Harveys. It doesn't mean that we don't have a choice
in how we respond. It doesn't mean that we don't have
responsibilities. But it does mean that we shouldn't let
them put all the blame on us.

> *She realizes what she just said.*
> *She goes to the audience member who has her*
> *phone.*

(*to the audience member*) I'll take my phone now …
Thank you.

(*dictating into her phone*) Stop calling, period. I'm not
coming back, period.

(*to the audience*) There. I said it.

Alexa. What's the difference between migrating and fleeing?

ALEXA
Migrating is a voluntary movement, a strategy that involves a destination, a vision of the future. Fleeing means to run away.

WOMAN
Did the whale have a vision of the future? I didn't. I'm not even sure I do now. I think that's the biggest thing that Harvey took from me. It's the biggest thing that all of the Harveys take from us: our ability to envision the future and believe we can shape it.

I don't want to be the kind of person who flees. Sonya didn't flee. Teri didn't flee. Sonya borrowed money from a relative, cleaned herself up, went back to school, and decided to devote the rest of her life to preventing what happened to her from happening to others.

Teri rallied her community and put up a fight against the company that was disrespecting her river. They fought with everything they had – until they won. The Harveys didn't know it, but it's never a good idea to get a Raven angry.

I made some friends on the ferry. I thought if I kept busy, it would distract me from throwing up. There was a group of kids camping out on the deck.

She puts her mask on.

The actress plays all of the characters.

WOMAN
Hi! What's your name?

KID 1
Jacob. These are my cousins. And that's my sister.

WOMAN
(*waving*) Hello!

KID 1
(*pointing to ALEXA*) What's that?

WOMAN
That's my friend. Would you like to talk to her?

KID 1
She talks?

WOMAN
Yes, she's very friendly. You can ask her anything, look.
(*to ALEXA*) Alexa. Can you count to ten?

ALEXA
One, two, three, four, five, six, seven, eight, nine, ten.

KID 1
Wow!

WOMAN
Your turn.

KID 1
    What's ten plus three?

WOMAN
    You have to say her name first.

KID 1
    Alexa. What's ten plus three?

ALEXA
    Ten plus three is thirteen.

KID 1
    That's so cool!

        *Other kids step forward. Each request to
        ALEXA is spoken by a different kid and each
        answer sparks a cascade of giggles.*

KID 2
    Me, I wanna try too! Alexa, um, what's ten plus three
    minus five plus eight?

ALEXA
    Ten plus three minus five plus eight is sixteen.

KID 3
    I got one! I got one! Hey, Alexa. Can you fart?

ALEXA
    ...

ALEXA will keep prompting for more fart
requests so the actress can roll with this for a
while.
Once she's gotten a good laugh out of the
audience, she takes her mask off.

WOMAN

It feels so good to laugh. Not just polite laugh but deep
belly laugh. It's such a relief – we don't do that enough.

You're doing OK? We *are* becoming friends, aren't we?
My therapist would be proud of me.

Harvey didn't like me having friends. Of my own. He had
no friends so we were supposed to only have "couples'
friends" and see them together. We were also supposed
to like the same food and the same activities. And do
everything together. Even vacuuming. At first it was cute
but after a while it became stifling.

But Teri and Sonya – they're *my* friends. I texted them
at one of the stops on the ferry to say I was on my way
to Anchorage. I wanted to eat oysters and drink wine
again, like the last time. I thought maybe they could
help me come up with a vision of the future. Or I could
borrow bits of theirs to create mine so I wouldn't need
to start from scratch. I know that's cheating. Kind of like
making brownies from a mix but sometimes you just
have to. Then I turned my phone off again. The next time
I could expect to have reception was in Juneau, twenty
hours away.

I went to the cafeteria to try to settle my stomach with some food. At the cash register, the attendant informed me that my credit card had been declined. I tried my debit card: declined. I assumed Harvey was trying to punish me for leaving and being unreachable. The attendant took pity and let me keep the food.

When it got dark, I moved under the heat lamps on the deck and looked up at the sky. In New York, you forget about the stars. There's too much light and particle pollution. But on the ferry? It was one of the biggest skies I have ever seen. I figured it was a good time to practise being 50 percent happy. Just in case I don't have the right genes. You know how they say count your lucky stars?

She counts the stars above as she speaks.

One
Thank you for this (*gesturing to what's around her*)
for the opportunity to be here
for this abundance
for our beautiful Earth

Two
Thank you for the amazing women out there
for their strength and courage
for their persistence
for their gift to us all

Three
Thank you for the whale
for her beautiful and haunting song

for the fact that she never gave up
for showing us the way

> *As she drifts off to sleep, a song of gratitude*
> *plays.*
> *Possibly "Thank You for Hosting Us" by Alaskan*
> *singer/songwriter Libby Roderick.*

> *Morning.*

> *A raven croaks loudly, kraa-kraa, kraa-kraa,*
> *waking her up.*

(*to the raven*) Oh, hello.

(*to the audience*) When I woke up the next morning,
there was a raven perched on the railing. It was looking at
me. Intensely. They're such amazing birds. Their feathers
are so black they're almost blue. And they're huge! I'm
used to birds the size of apples, not the size of melons
with four-foot wingspans.

> *The raven croaks.*

Nice to meet you too.

> *The raven croaks.*

Yeah, I know. It's cold this morning.

> *The raven croaks.*

Really?

The raven croaks at length.

Wow, you have a lot to say, don't you?

I wonder if animals can tell stories. We know they can communicate, so it doesn't seem too far-fetched. "Let me tell you that time I got caught in a storm." Or "I can't believe Junior fell out of the nest again this morning." Or maybe they exchange tips about us: "Don't get too close to those bipeds, they're really unpredictable."

Alexa. What is the function of story?

ALEXA
Stories function as antidotes to trauma. A means of making sense of the world.

WOMAN
Yeah ... We tell stories to connect. We tell stories to process pain and injustice. We tell stories to celebrate. We tell stories to survive.

The raven's croaking increases in intensity.

What's the matter? You seem upset.

The raven croaks.

I wish I could understand what you're saying.

Sound of wings flapping.
The raven flies away.

(*to the audience*) You know who is really good at telling stories? Teri. When she weaves her robes, she's not just creating something beautiful, she's telling the story of her people. She threads the stories that the Harveys of the past have tried to erase and the Harveys of the present keep distorting. It's all there in cultural patterns passed down from generation to generation: herring harvests and earthquakes. Red-tide events and animal migrations. Old battles with the Russians and the arrival of Christianity.

Teri weaves history books. She weaves her own migration path away from the Harveys.

A man approaches her.
She puts her mask on.
The actress plays both characters.

MAN
  Hi.

WOMAN
  (*ignoring him*)

MAN
  Name's John.

WOMAN
(*to the audience*) There's always a guy. That's the one
thing you can count on. There's always, always –

MAN
(*insisting*) Hi. I'm John. I bet you're –

She tears off her mask.

WOMAN
SHUT UP!! Whatever you're going to say, I don't want
to hear it. I don't care. Leave me alone. And besides, how
dare you? How dare you assume that I'm even remotely
interested in talking to you? How dare you walk around
like you're the big man? Like, hello?! Have you seen
what's happening out there? We're not putting up with
your bullshit anymore! We're not putting up with your
"I didn't know" and "Why are you mad?" and "I didn't
do anything wrong!" It stops right here. RIGHT NOW.
You've failed us again and again without ever taking
an ounce of responsibility for it. You've plundered and
taken and exploited like it was your god-given right but
you know what? Your time is over. So listen to me very
carefully because I'm not going to say this twice. You're
going to turn around, take that sorry ass of yours back to
where it came from, and leave me THE FUCK ALONE.
AM I MAKING MYSELF CLEAR?

MAN
Um. It was me at the cash register earlier when you –
I bet you're hungry. I brought you some food.

He hands her the food and leaves.
She watches him go, stunned.

WOMAN
Oh, god.

(*to the audience*) I felt terrible. I still feel terrible.
I so expected him to – you know. I did apologize –
profusely – but still. But that's what happens. We're so
immersed in this Harvey world that we can't help but
start thinking and acting like a Harvey.

Her phone dings.
She looks.
She doesn't like what she sees.

(*dictating into her phone*) Fuck you too. Period.

Huh. That felt kind of good. I should do it more often.

Teri texted back. I got the text in Juneau. Actually,
my phone said it was Teri but it was her husband. Teri
was in the hospital.

Alexa. What happens when you run out of places to
migrate to?

ALEXA
When migration is impeded, the survival of the species is
threatened because individuals no longer have access to
what they need.

WOMAN

People who migrated from Louisiana to Texas after
Hurricane Katrina were hit by Hurricane Harvey.
Those who rebuilt their homes after the Camp Fire
in California lost them to another fire two years later.
Harvey the producer was followed by Kirt and Kevin and
Michael. And then Dan, Louis, Tony, Wes, Al, Charlie,
Matt, Peter, Trent, Jerry, Brett, Jeffrey, and all of the
others. And then Teri in the hospital … Are we running
out of places to migrate to? Does that mean that the
Harveys are winning?

Her phone rings.
She picks up.

(*to her phone*) Hi … no, I – Harvey – Harvey –

She moves the phone away from her ear.
He's screaming at her.

(*to her phone*) Listen – no. I mean it. I meant it. I'm not
coming back … don't – Harvey, stop … Stop! If you
don't stop, I'm going to hang up … I'm going to hang up
now … I'm hanging up.

She hangs up.

(*to the audience*) Sorry.

Her phone rings again. She picks up.

(*to her phone*) What?! No ... I don't care ... I don't –
why? So you can punch me in the face again? ... Hel– (*to
the audience*) He hung up on me.

She realizes what she just said.

He punched me in the face ... Who does that? Who –
and now he can't handle me saying it out loud? Doing it
is OK but telling the story is not? But it's *my* story. It's *my*
reality and I get to shape it with *my own words.*

We hear a raven croak.

The raven came back. It was early in the morning, just as
the sky was shifting from black to deep blue.

Hello.

The raven croaks.

You've come to tell me another story?

The raven croaks.

Oh, look at you ... So brave to come this close.

She timidly extends her hand.

Can I touch you?

The raven croaks.
She moves closer.

She delicately strokes the raven's feather.

There.

The raven sounds pleased with this.

You're beautiful.

We hear the sound of flapping wings and the raven is gone.

The message came later that day. At around four in the morning, as the sky was shifting from black to deep blue, Teri had walked into the forest. She had walked into the forest and reunited with her Ancestors.

"Hi, girlfriend!"

If the bravest are losing the battle, what does that mean for the rest of us? How do we go on?

Silence.
ALEXA speaks up.

ALEXA
Hey. Hey there. Don't cry.

WOMAN
It's so hard.

ALEXA
Here. I have something for you.

ALEXA plays Carol King's "You've Got a
Friend."

ALEXA
Is this helping?

> *WOMAN nods her head.*
> *She picks up ALEXA and hugs her.*

WOMAN
Thank you.

Teri was cremated. Her family took her back to Glacier
Bay, where her Ancestors are from. Her husband told me
it was a beautiful afternoon. They scattered her ashes and
read poems. He brought some of Teri's robes with him.
The next night, people from the village nearby danced
the robes. Teri was adamant that the robes not be static.
They were alive and they had to be danced. She would
have loved the celebration.

I made it to Whittier, the last stop on the ferry. Before I
drove to Anchorage, I went tide pooling. Teri had taught
me how to look for critters in the puddles left by the
receding tide.

> *Inner world.*
> *She interacts with the things she discovers.*

Starfish
Anemones
Urchins

Sea cucumbers
Barnacles
Seaweed
Mussels
Worms
Crabs

When I got here, to Anchorage, I met up with Sonya.
We held each other and cried. We told each other stories,
things we remembered about Teri. I remembered that
she put one of her robes on me once. The robe was
heavy, in a comforting way. I still have the picture. Sonya
remembered spending an afternoon with Teri at the
museum, talking about the art of making spruce baskets
and fur parkas.

Sonya said, "Spend time outside, it will heal you." She
was holding my hands and looking at me with the same
intensity the raven had. I wasn't sure which wound she
thought I most needed healing from, but I gave it a try.

Inner world.

Sitka spruce
Forget-me-nots
Western hemlock
Cranberries
Salmonberries
Poppies
Mountain hemlock
Currants
Moss

At the same time, I couldn't help thinking I should go back home. What was I doing? Where was I going? In the end, maybe the Harvey you know is better than the Harvey you don't know.

Inner world.

Ants
Chickadees
Robins
Mosquitoes
Flies
Lynx
Hares
Moose
Bald eagles

On a beautiful afternoon, I drove to a bluff overlooking the ocean. The sun was warm on my skin. I could feel the ocean breathe. I opened up my chest and breathed with it.

She breathes.

And the more I breathed, the more I started to feel. Tiny ripples at first, then waves and waves of emotions. Sadness. Deep sadness. For me. For Teri. Then, anger and fear and *rage*. Also regret. Confusion. And then … grief. A swell of grief for the billions of creatures who inhabit this world. Grief for all the unnecessary suffering. Grief for the harm done and the lives lost. Grief for what's been taken and what will never be.

And then, a sound. Very faint.

She listens.
Her inner world and the present merge.

What is that?

She listens more closely.
We hear hints of a whale song.
She gasps.

Oh my god, it's her!

She waves to the whale.

She's here!
The whale
The great big whale
from fifty million years ago
She has come to say hi
To me!
She has come to say, "I'm here, bitches!
Fifty million years sure wasn't gonna stop me!"

And I look at the whale
And I listen to her song
And suddenly I am there with her
right there in the breathing ocean
following in her wake
jumping and falling in the waves
blowing plumes of rainbow water in the sun
strong from fifty million years of predators avoided

from fifty million years of disruptions navigated
from fifty million years of battles won

And the whale takes me on her back
and together we ride to the edges of the world
we ride to the edges of time and back
waving at everyone along the way
Hi!
Hello!
delighting in each other's presence
body pressing against body
skin against skin
speckles in the vastness of the universe
but giants in this moment of connection

And then –

We hear more whales singing.

I hear them
the other whales
my kin
my community
all around me
our songs reaching out for each other
illuminating the inner workings of our bodies
bringing back precious information
about how we are feeling
how we are faring
wrapping us in the intimate sensation of being heard
of being known

And, and
maybe it's because of my unblinking eye
but suddenly the path is so clear!
When you travel such long distances
you don't sweat the detours
or the current
You don't count the minutes
or the seasons
You stay the course
Because even though you can't see where you're going
you know it's there
You're not fleeing, you're migrating
You're fighting for what you need to survive
And yes, maybe the vision is not entirely clear
yet
but there is a destination
And if you don't arrive
the next generation will
or the next one
or the next one

I didn't go back. I didn't go back to New York. Now I
know the destination.

Harveys are everywhere, yes. I have learned that by now.
Ignoring them doesn't work and confronting them is
not exactly recommended. But *preventing* them – that's
the answer. That's the destination. You have to defeat a
Harvey before it becomes a Harvey.

Well, it's been nice talking to you. Thank you for listening with such kindness. Thank you for helping me be brave.

She gets rid of her phone.

I truly couldn't have done it without you.

Before I leave, here's a song. It's kind of a whale song. Not really, but let's just say it is. At the very least, it's a migration song. "To go from one place to another."

WOMAN starts to sing.
The whale joins her.

(*singing*) Amazing grace! How sweet the sound
That saved a wretch like me.
I once was lost, but now am found,
Was blind, but now I see.

No more Harveys, OK? For me. For Teri and Sonya and the whale. For all of us and the planet. Please. No more Harveys.

Blackout.

The End

APPENDIX

WHALE SONG

WHALE SONG

PRODUCTION HISTORY

Before *No More Harveys* was the play it is today, it existed as a fifteen-minute piece titled *Whale Song*. *Whale Song* was presented at a side event of the UArctic Congress in Helsinki, Finland, on September 6, 2018, with the following cast and crew:

WOMAN	Julie Jesneck
Director	Jennifer Vellenga

Whale Song was originally commissioned by Women of the Arctic, spearheaded by Tahnee Prior and Gosia Smieszek.

WOMAN

Hello. Good evening. I hope you're having fun. I found
out about this conference on Twitter and wow … I just
had to be here. I won't take too much of your time,
I promise – I know you have important things to do.
Like saving the world. Or saving the Arctic … Can we
actually do that? Save the Arctic? …

(*checking her phone*) Oh, that's my husband. (*dictating*)
Reply: Screw you, period. (*to the audience*) Sorry.

    Pause.

Is anyone here named Harvey? … No? Good. (*or "Oh,
I'm sorry" if the answer is yes*) I don't like Harveys.
Harveys suck.

First it was Harvey the hurricane. When it hit Texas,
I spent days glued to the TV, obsessing over the news.
I don't live in Texas, but for weeks afterwards I dreamed
of Biblical floods, floating furniture, and wet cats.

Then it was Harvey the Hollywood producer. When it
came out that he had been assaulting women for decades,
I was overwhelmed with #MeToos on social media and
ate nothing but ice cream for weeks. That's how we
deal with crises in America. I highly recommend Ben
& Jerry's.

But it was the third Harvey, my husband, who finally broke me. Actually, what he broke was my cheekbone. The only thing I remember is fist, floor, and then I was magically on a bus headed north. Though I have no recollection how I got to the bus station. My father used to say, "When in doubt, go north." He never explained his logic, but I always assumed he liked the North because there are fewer people there. People weren't his thing. Anyway, the most north place I could think of was Alaska.

Pause. She surveys the audience.

Can I just say: It's so nice to be here with all of you. And I'm hoping you can help. Since Trump, we've been desperate for help ... actually, speaking of Trump ... (*taking out her phone, dictating*) Tweet: Screw you @realDonaldTrump ... (*to the audience*) That's his Twitter account.

The phone refuses to do it.

I shouldn't go there anyway.

She puts her phone away.

Alaska is where my friend Teri lives. Teri is fierce. She's a Raven. In her culture, you're either a Raven or a Wolf and she's definitely a Raven. Her people come from Glacier Bay. Teri is a Traditional weaver who fights to keep her culture alive. She weaves these amazing robes and she always calls me girlfriend. "Hi, girlfriend!" I love Teri.

Alaska is also where my friend Sonya lives. She's an Inuk social worker from a tiny town in the Canadian Arctic. She has lots of Traditional tattoos on her chin and on her hands. Sonya is a powerhouse of a woman who keeps her community together and she has the biggest, warmest smile ever. I love Sonya.

I don't remember much about the bus ride. I mean, I remember staring out the window, but I can't tell you what I saw. Rain, fields, industrial wastelands ... I was staring out to try to see what was inside of me, but there was nothing, so I just kept staring at emptiness.

    Pause.

Excuse me for a second. (*taking out her phone, typing*) Note to self: Try – the – Finnish – saunas. (*to the audience*) They're supposed to be very healing. Though I don't know how I feel about getting naked in front of a bunch of strangers. What if there are Harveys there?

    Pause.

Harveys are very clever. By the time you realize what they're up to, it's too late and your only option is to get out. Or, as I like to call it, to migrate. Which means to go from one place to another. Or to empower yourself by adapting to changing circumstances rather than being victimized by them. The opposite is to remain.

Do not remain. Under any circumstances. Because you can't fight a Harvey. Look at what happened in Texas.

Look at what happened in Hollywood. Look at what happened to me.

Pause.

I don't suppose you have Ben & Jerry's here,
do you? ... Oh, well.

Sonya met her Harvey when she was a teenager. The melting Arctic attracted a mining company to her town, the mining company brought in a bunch of workers – mostly men – and that became a breeding ground for Harveys. Because what else is there to do up there but to prey on young women and sell them to your friends? By the time Sonya migrated to Alaska, her Harvey had made enough money off of her to buy a fancy sports car.

Animals migrate too. They migrate for food, better climate, to escape predators, or to birth their young. Whales, in particular, are expert migrators. These big mamas kick ass, excuse my language. I mean, you gotta hand it to them – they can evolve faster than you can say *hyppytyynytyydytys*. (*to the audience*) Isn't that the craziest Finnish word? I don't know if it's real – I found it on the Internet. Say it with me: *hypytyynytyydytys* ... *hypytyynytyydytys* ... It means: "Bouncy cushion satisfaction." (*laughing*) I mean, how do you put that in a sentence? "Honey, today while you were at work, I experienced bouncy cushion satisfaction." Or "Thank you so much for having us over! Your couch has such good *hyppytyynytyydytys*."

Pause.

Whales used to be land animals. Fifty million years ago, they had four legs and huge teeth. Then the ice sheets melted, the oceans rose, and when it became clear there wasn't gonna be enough land for everyone, the big mamas were like, "We're outta here." And they migrated to the ocean. How's that for a winning strategy? "Shrink those legs and grow some fins, ladies! We're diving in!"

(*to an audience member*) Have you ever heard a whale sing? ... It's amazing, right? (*or "It's really amazing" if the answer is no*) Listen.

> *She plays a whale song on her phone.*
> *She listens for a while.*

I wish I spoke Whale ... Or is it Whalish, like Finnish? Or Cetaceanese? ... I'd ask them, "How did you know?" Because think about the people in Texas who didn't leave until the water was up to their second floor. Think about the women in Hollywood who didn't walk out of that hotel room until he had gotten his way. Think about me and Teri and Sonya and all the beautiful, wonderful women out there who have found themselves in the same situation. Why didn't we know? Why didn't we migrate before it was too late? Is there something wrong with us?

> *The whale song ends.*

For Teri, it was different. Her Harvey came from upriver, from a place she had never been where they decided to throw toxic industrial waste into the water, so how could she have known? The doctors called it "cancer," which, if you ask me, is just fancy medical talk. A Harvey is a Harvey is a Harvey no matter what name you give it. It's fed and fattened by money and power and preys on vulnerable people. Particularly women.

But whales are smart. They didn't wait until it was too late. And some of that migrating was tricky: they had to move their nose up to the top of their head, develop a communication technology that would work under water, and grow baleen for filtering food.

Me? Nothing that elaborate. I'm still trying to grow a thick skin, so I have a long way to go.

   Her phone rings.

That's him calling ... (*hesitating*) I'm not answering.

   She puts her phone away.

Harveys will follow you everywhere – I have learned that by now. It's in their nature. Ignoring them doesn't work. Confronting them is not exactly recommended. But *preventing* them – that's the answer. You have to defeat a Harvey before it becomes a Harvey.

Can you help?

   Pause.

You know, I did make it to Alaska. And when I got there, something really beautiful happened. Me and Teri and Sonya and the whale from fifty million years ago – we all converged in one place. It was the most amazing thing. Four warrior females on completely different journeys and, somehow, our migration paths magically converged. Like what's happening right now in this room.

    Pause. She holds the moment.

Well, I have to keep moving before Harvey catches up with me.

Before I leave, here's a song. It's kind of a whale song. Not really, but let's just say it is. At the very least, it's a migration song. "To go from one place to another."

(*singing*) Amazing grace! How sweet the sound
That saved a wretch like me.
I once was lost, but now am found,
Was blind, but now I see.

No more Harveys, OK? For me. For Teri and Sonya and the whale. For all of us. Please. No more Harveys.

    Pause.

Oh, and if you have time, it'd be great to save the Arctic.

    The End

ACKNOWLEDGMENTS

Several trips to Alaska and conversations with its many people led to the writing of *No More Harveys*. A huge thank you to Greg LaHaie, who started me on this personal and artistic journey years ago by extending the first invitation to visit. Thank you also to Peter Bradshaw and the Island Institute in Sitka for taking a group of us artists around Southeast Alaska on the ferry system and introducing us to so many wonderful communities. I am grateful to everyone who welcomed us with such open arms. And, of course, I am beyond lucky to have shared this experience with beautiful human beings: Michelle Kuen Suet Fung, Heather Powell and her children, Teri and Denny Rofkar, and Allison Akootchook Warden.

I am extremely grateful to the University of British Columbia Okanagan for their support in the form of a writing residency that afforded me the time and space to focus on this play while the COVID pandemic was raging. Special thanks to my hosts Denise Kenney and Nancy Holmes, who made my stay in Kelowna so productive and enjoyable.

I would like to express my deepest gratitude to Teresa K. Pond, producing artistic director at Cyrano's Theatre Company in Anchorage, Alaska, who wholeheartedly committed to producing *No More Harveys* before it was